THE LITTLE BOOK OF
SACRED
GEOMETRY

THE LITTLE BOOK OF SACRED GEOMETRY

Text by David Olliff

An Hachette UK Company
www.hachette.co.uk

Summersdale Publishers Ltd
Part of Octopus Publishing Group Limited
Carmelite House
50 Victoria Embankment
LONDON
EC4Y 0DZ
UK

www.summersdale.com

Printed and bound in China

ISBN: 978-1-80007-682-2

Substantial discounts on bulk quantities of Summersdale books are available to corporations, professional associations and other organizations. For details contact general enquiries: telephone: +44 (0) 1243 771107 or email: enquiries@summersdale.com.

DISCLAIMER
The author and the publisher cannot accept responsibility for any misuse or misunderstanding of any information contained herein, or any loss, damage or injury, be it health, financial or otherwise, suffered by any individual or group acting upon or relying on information contained herein.

THE LITTLE BOOK OF
SACRED
GEOMETRY

ASTRID CARVEL

summersdale

॰ CONTENTS ॰

◌ **INTRODUCTION** ◌

"To see a world in a grain of sand, And a heaven in a wildflower." When William Blake wrote these words, he was inviting us to take time, to meditate on the eternal beauty of things, to be mindful, and to allow ourselves to be inspired. Simply noticing the beauty, structure and patterns of creation is to connect with the sacred. Today it seems we all need a reminder to notice things, to reconnect and to realize our connectedness.

This book is an invitation to take a journey towards connection. Along the way, we will discover shapes, numbers and patterns woven into the fabric of nature. We will uncover the mathematical properties of beauty and the mystical side of geometry. We will reveal fascinating relationships with shapes and numbers – and find new ways to see the sacred.

We will take a look at how our ancestors understood the divine nature of number and form, and how they encoded it into sacred designs, art and architecture.

Finally, you will find some suggestions on how to bring the renewing power of sacred geometry into your spiritual life. Welcome to a new way of seeing the connectedness of being.

To get the most out of this book, you will need:

- A calculator
- Colouring pens or pencils
- A compass
- Crystals
- A glue stick
- Graph paper
- A journal
- A ruler
- A protractor

WHAT IS SACRED GEOMETRY?

Have you ever had the feeling that the universe is trying to tell you something? In everyday life, we don't always notice the wonder of it all. This world, this universe, is bursting with life and creative energy. New stars are forming and reforming in swirling galaxies, while new life is emerging, growing and renewing all around us here on Earth. The same creative impulse that shapes galaxies reveals itself in the most intricate and structured patterns of nature. These patterns are one of the languages of life; while it writes in numbers, it speaks to the heart.

In this chapter, we will learn more about this language of life. Prepare to count flower petals and to find ratios and irrational numbers. Get ready to meet the most beautiful rectangle ever and a truly sublime triangle. This is geometry, but not as you might remember it!

ENDLESS FORMS MOST BEAUTIFUL
AND MOST WONDERFUL HAVE
BEEN, AND ARE BEING, EVOLVED.

CHARLES DARWIN

ͼ **MEASURING THE EARTH** ͽ

Geometry literally means earth measurement. The first part of the word comes from the Greek for earth, which is related to the name Gaia, the Earth mother. The second part comes from the Greek to measure. Geometry may well have been something you did in a maths class, but in antiquity, it was part of the philosophical quest to understand the whole world, and something of the realm beyond. When eternal truths are found, they are as true for the material world as they are for the realm of the gods.

For the ancient philosophers, a term like sacred geometry would have been an unnecessary repetition. To them, geometry could not be anything other than sacred. The universe and human beings were reflections of the eternal and divine; all things were connected. The modern scientific mindset has led us to think of geometry as purely logical for analysing problems. Reconnecting geometry with the divine reminds us to treat all things, ourselves included, as beautifully designed and wonderfully sacred.

❦ THE INTRICATE WEB OF LIFE ❦

Take a look at a spider's web. Here is a pattern so intricate, balanced and symmetrical it seems almost impossible that a tiny creature could have created it by simple instinct. And yet, these incredibly strong and durable structures are encoded into the very nature of spiders. Their webs may be beautiful, but they are also essential for their survival. From the spider's point of view, these webs are simply nets that they use to catch lunch.

Sacred geometry is all about looking at nature with fresh eyes. There are structures everywhere which depend on patterns of symmetry and the principles of number and geometry for strength, growth, survival and beauty. Discovering the underlying geometry of life brings an awareness of the primal energy that first created it. Whether its source is God, Goddess, or the inherent life force of qi, this creative energy is divine in nature and sacred to life. All things are connected by it. This energy, this source of divine abundance, seems to flow most fruitfully through the patterns and shapes familiar to geometry.

TIME IS A SPIRAL — THE CYCLES
ENDLESSLY REPEATING, YET
ALWAYS MOVING. KNOW THE
SPIRAL AS THE UNDERLYING
FORM OF ALL ENERGY.

STARHAWK

⌒ **SACRED GEOMETRY IN NATURE** ⌒

If you learned geometry at school, the subject might seem to have very little to do with the natural world. The truth is, however, that there are many principles of number and proportion, line and circle, form and structure that are as easily found in the garden as they are in a textbook. Imagine for a moment a garden bursting with life in the summer, with daisies in the lawn. Those common daisies, little flowers with yellow circular centres surrounded by thin white petals, are a beautiful encounter with sacred geometry. The central dome is made up of dozens of tiny disc florets, packed together in an intricate spiral pattern, while a specific number of white petals radiate out. The spiral in the centre, and the number of petals that surround it, follow a geometric proportion and a numeric sequence that is found in many of nature's most beautiful structures. These flowers, and a great many others in the garden, are living geometric patterns expressed by a sequence of numbers called the Fibonacci sequence.

THE RABBITS IN FIBONACCI'S FIELD

Suppose a man owns a field; let's call him Fibonacci. One day a pair of rabbits appear in the field. For one month, they live together: January, one pair. By the next month, rabbits being rabbits, one of the pair is expecting babies: February, still one pair. By the next month, a pair of baby rabbits are born: March, two pairs of rabbits living in Fibonacci's field. By the next month, another pair are born: April, three pairs. One month later and two more pairs are born: May, five pairs.

This pattern of growth goes on, with each pair of rabbits reproducing after one month and then giving birth to two rabbits after another month: June, eight pairs. July, 13 pairs. August, 21 pairs. September, 34. October, 55. November, 89. Then, finally December – just one year later – and there are 144 pairs of rabbits. This is the Fibonacci sequence: starting with zero, each number is the sum of the previous two: 0, 1, 1, 2, 3, 5, 8, 13, 21, 34, 55, 89, 144...

FIBONACCI'S NUMBERS AND THE DAISIES IN THE LAWN

"He loves me. He loves me not. He loves me. He loves me not..." Someone idly pulling little white petals from a common lawn daisy on a lazy afternoon will likely be unaware of their encounter with the Fibonacci sequence. All things being equal, there will be 34 or 55 white petals scattered to the wind – both are Fibonacci numbers.

A great many flowers follow the Fibonacci sequence in this way. A peace lily has one petal, the iris has three. A buttercup has five petals, while delphiniums have eight. A black-eyed Susan has 13 petals and a Shasta daisy 21. Rose petals are a beautiful example of the Fibonacci sequence in nature. Each new set of petals grows in the spaces of the previous. This arrangement creates a spiral of layers in which a new petal is the sum of two previous ones. It is a highly efficient arrangement, which creates those tightly packed blooms, ensuring effective distribution of sunlight to the whole flower.

The Fibonacci sequence is named after the Italian mathematician who wrote about it in the thirteenth century. It was Fibonacci who used the example of rabbit populations to think about a pattern of growth in which a number always depends on the previous two. As the values increase, they do so in regular proportion to each other. This pattern of growth is routinely found in things that grow. Most plants exhibit some characteristics of the Fibonacci sequence and the Golden Ratio. The seeds on a sunflower are packed onto the seed head in the most efficient arrangement possible because they grow in a spiral pattern made by tiny rotations based on the Golden Ratio. The number four, which is not a Fibonacci number, is rare when it comes to plants. Maybe this is why four-leaf clovers are so special.

Next time you find yourself in a garden, try some counting. Count petals, the leaves on a stem or the seeds which make a curve on a sunflower head. You will find Fibonacci numbers everywhere!

THE FIBONACCI SEQUENCE AND THE GOLDEN RATIO

The Fibonacci sequence continues to fascinate mathematicians because the numbers in the sequence relate to each other in often remarkable ways. To explore these very special numbers, we can start with the Golden Ratio.

Mathematically speaking, the Golden Ratio is a decimal number which has been assigned the Greek letter Phi. In so far as it is a geometric ratio which is infinite, Phi (φ) is similar to another number assigned to a Greek letter, Pi (π). The value of Phi is given as 1.618033... and so on into infinity.

To see how this ratio is revealed within the Fibonacci sequence, we can work with some numbers in the early part of the sequence. Let's start with 3, 5 and 8. To see the Golden Ratio, take the larger number (the sum of 3 and 5) and use your calculator to divide by the previous number:

$$8 \div 5 = 1.6$$

This is pretty close to the Golden Ratio, but then, we are only a few numbers into the sequence. Let's look at the numbers 55, 89 and 144:

$$144 \div 89 = 1.617977...$$

This is much closer to the Golden Ratio already, and these numbers are only 12 and 13 into the series (if you start at zero).

Now for some more intriguing features of these special numbers. Remember, the Fibonacci sequence reveals growth, and concerns the results of the two that have gone before. For this reason, it is good to look at the numbers in threes.

Back to 3, 5 and 8. Let's divide the numbers next to a number by the number in the middle (3 divided by 5, then 8 divided by 5):

$$3 \div 5 = 0.6 \qquad 8 \div 5 = 1.6$$

The difference between the two results is always exactly one, all the way through the Fibonacci sequence. Now let's try something else on the numbers surrounding 5. Multiply the two surrounding numbers then multiply the middle number by itself:

$$3 \times 8 = 24 \qquad 5 \times 5 = 25$$

The middle number squared is always exactly one more than the surrounding numbers multiplied. This is true all the way up the sequence. Search online to find the first 100 Fibonacci numbers and try it with three Fibonacci numbers over one million!

TO SEE A WORLD IN A GRAIN OF SAND
AND A HEAVEN IN A WILD FLOWER,

HOLD INFINITY IN THE
PALM OF YOUR HAND

AND ETERNITY IN AN HOUR.

WILLIAM BLAKE

POSTCARDS AND OTHER PLEASING RECTANGLES

One of the curious things about the Golden Ratio is the way in which it seems to be associated with the geometric shapes that we find most pleasing. Long before the Fibonacci sequence was found to conform to the Golden Ratio, the ratio was realized to be a balanced proportion – this somehow just felt right. Perhaps this was the reason why the Golden Ratio became known as the divine proportion in the 1500s. Might this ratio have been woven into the fabric of creation by God?

The Golden Ratio may be explored in geometry through rectangles, and the proportions of one rectangle, known as the golden rectangle. This rectangle has sides in proportion to the Golden Ratio, and it happens to be one we generally find most pleasing. Postcards are often golden rectangles, as are bank cards.

Golden rectangles can give shape to the entire Fibonacci sequence, and as they do, they reveal the very spirals that are found in nature. The fact rectangles can reveal spirals is just one of the curious wonders of sacred geometry.

THE GOLDEN RECTANGLE AND THE GROWING SPIRAL

Using graph paper, draw two rectangles with sides of one unit. These first rectangles are squares, of course. Placed next to each other, they make an outer rectangle, which is 1 by 2 – both Fibonacci numbers:

Now add a rectangle (square) with sides that match the longest side of your 1 by 2 outer rectangle. The resulting outer rectangle is 2 by 3 – Fibonacci numbers in the ratio 1.5 (heading towards the Golden Ratio):

Now add a rectangle (square) with sides that match the longest side of the outer rectangle to form a new rectangle, 3 by 5, which is in the ratio 1.666... Then add another. This gives an outer rectangle of 5 by 8, which is in the ratio 1.6. This growth pattern can go on and on:

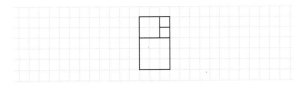

Each of these outer rectangles is more and more closely aligned with the Golden Ratio, so each is also in closer proportion to the golden rectangle. The proportion can never be exact, of course, as the ratio itself is infinite. This is a bit like trying to form a perfect circle, knowing that the circle ratio Pi (3.1415927...) is infinite. Now all you have to do is join the points across each square with a curve. As you do this, a spiral is revealed, which expands outward on the proportions of the Golden Ratio, following the Fibonacci sequence:

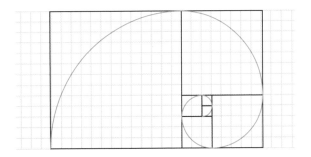

⌒ **THE GOLDEN TRIANGLE** ⌒

Perhaps unsurprisingly, there is also a golden triangle. A golden triangle, which is also rather beautifully known as a sublime triangle, is an isosceles triangle where the base length is in a golden ratio with the side length. So, if you multiply a base length by the Golden Ratio, you will find the side length. One angle of a golden triangle will equal 36°, leaving the other two angles at 72°.

There is a closely related triangle to this one, which has the lovely name of golden gnomon. The golden gnomon has one angle of 108° and two angles of 36° each. For the golden gnomon, the side length is in a golden ratio with the base length.

Take a look at the nested triangles below. A golden gnomon (the flatter triangle) placed next to a golden triangle (the taller triangle) forms another golden triangle giving the points for our familiar spiral.

THE FUNNY THING ABOUT FIVE

At this point, it's worth drawing attention to the number five. It turns up a lot in sacred geometry. We will see how the pentagon with its five angles is made up of both kinds of golden triangle, even into infinity. There is the pentagram with five points made up of five lines (see pages 28–29). We can find plenty of fives in human anatomy: five fingers, five toes, five openings to the face. Five, of course, is a Fibonacci number.

As we approach the sacred shapes and align them with the elements, again you will find there are five. Fire, water, air and earth are material elements, but there is a fifth one that is known as quintessence, which literally means fifth essence. This is identified as spirit.

Five is significant in faith too. Muslims follow the five pillars of Islam and pray five times a day. Christians remember the five wounds of Christ.

⌒ **THE SPIRAL DANCE** ⌒

As forest ferns develop new growth in the spring, their leaves expand by unrolling tight spirals known as fiddleheads. The pinecones that scatter the forest floor in the autumn are made up of scales that protect the seed in a double-spiral pattern, based on adjacent Fibonacci numbers; 13 spirals clockwise or eight spirals counterclockwise. Those seeds in a sunflower arrange themselves similarly in a double-spiral pattern based on adjacent Fibonacci numbers; 55 spirals clockwise or 34 spirals counterclockwise.

Fibonacci numbers make a particular kind of spiral that expands outwards. This is known as a logarithmic spiral in the way that it grows out as it turns. The Fibonacci sequence is a pattern of expansive growth, which Fibonacci illustrated with his rabbit scenario (see page 15). The associated spiral is a geometric structure of expansive, organic growth which occurs naturally throughout the living world.

A famous example of this kind of expansive growth spiral is the shell of the nautilus mollusc. The chambers of the nautilus shell grow larger with each phase of renewal in a proportion which follows the Golden Ratio. This allows the nautilus mollusc to continually grow and yet still inhabit the same shell.

There is beauty in the spirals of flower seeds or shells, but we should remember these spirals are also a matter of survival, protection, efficiency and renewal. The seeds in the sunflower grow this way to pack as many seeds as possible onto the seed head, with the most efficient use of energy. More seeds on a seedhead mean more sunflowers next summer.

Sacred geometry invites us to find the divine pattern in the everyday. It also invites us to realize the divine pattern in ourselves. The spiral inspires us to recognize our infinite potential for growth and renewal. When we accept this divine abundance, we realize our growth is unlimited, and we can allow ourselves to live in harmony with life.

⌒ **THE PENTAGRAM** ⌒

Perhaps one of the most widely recognized shapes in many fields of spirituality is the pentagram. It is sacred to neopagans and Wiccans and central to ritual magic. Of course, its use and symbolism go back much further. The Pythagoreans understood it as a symbol of health some 2,500 years ago. It also decorated the shield of Gawain in the fourteenth-century poem, *Sir Gawain and the Green Knight*.

The meaning of the pentagram centres around the number five: five senses, five points of the human body, five elements (fire, water, air, earth and spirit). Five is a very important number when it comes to sacred geometry.

The pentagram is in every sense a golden geometric structure. The pentagram is in every sense a golden geometric structure. The five points are all golden triangles, while the inner pentagon invites a further pentagram turning into infinity. What's more, all five lines that make up the pentagram are in a golden ratio with intersects of each other.

The horizontal dark green line is proportionally longer than the lighter green line by the Golden Ratio. The dotted green line is proportionally longer than the grey line by the Golden Ratio.

Now take a look at the upward-pointing central triangle formed within the outer pentagon in the diagram below. This is a golden triangle. The two triangles either side of it that complete the pentagon are, wonderfully, golden gnomon triangles. In fact, in the diagram below, all the triangles are golden in either of these two ways. Try counting them (hint: there are more than 50). Surely perfection like this is rare? Well, geometrically, and even naturally, it's really not.

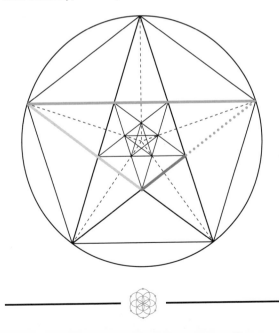

☌ **THE SACRED SPIRAL** ☌

We began this chapter imagining the shape of a spider's web, an extraordinary pattern and structure. It is formed in a tight spiral to a clear design of symmetry and strength. As we have progressed through the examples of organic patterns of growth, we have found them expressed in a numeric sequence, a ratio, and a spiral. These numbers and ratios have revealed themselves in plain geometric forms, from the rectangle to the triangle and the pentagram. All of these are encoded with the ratio that returns us to the spiral.

The spiral is the spring-like energy of emergence and the stored energy of renewal. Our planet orbits a sun on the outer edge of a vast spiral arm of stars that swirls around the centre of our galaxy. And then there is the unimaginably small: the double-helix DNA is a spiral structure that carries the instructions for life itself.

Our next chapter introduces the fundamental sacred shapes. Of all the sacred shapes, the spiral is the basic form of all creative energy: the sacred spiral of life.

THE SACRED
SHAPES

From the previous chapter, we have seen the patterns, proportions and numbers that seem to permeate the whole of creation. We have also seen geometric shapes that follow these divine proportions: golden rectangles, triangles and spirals as well as the sacred pentagram. In this chapter, we will encounter a full range of sacred geometric shapes.

Each of these sacred shapes carries its own spiritual significance. Each resonates with its own divine energy. All of the sacred shapes may be accessed through meditative practice which enables us to attune to their healing and rejuvenating power. The more we experience the sacred shapes in this way, the more they will raise us towards higher consciousness and deeper connectedness with the divine source of life.

⌒ **THE CIRCLE** ⌒

An intriguing question to ask about a circle is how many sides does it have? Maybe one side, or maybe infinite sides very nearly reaching oneness. The circles we draw can never be perfect, but they draw us towards the eternal idea of a perfect circle.

Circles in geometry depend on Pi. This is the ratio of the circumference to the diameter of the circle, which is just a little over three. Just like the Golden Ratio Phi, Pi can be calculated into infinity. This is one way to see circles as infinite, but they also represent eternity. Rings are circles, which have long been used to symbolize eternity, such as in the exchange of wedding rings or later in the giving of an eternity ring.

The circle can therefore represent wholeness, oneness, unity, perfection, and the never-ending cycle of life. The ancient symbol of the ouroboros, the serpent which eats its own tail, represents infinity, wholeness and immortality. It reveals a process of self-renewal and regeneration as the serpent consumes the old self in order to be reborn.

⌁ THE *VESICA PISCIS* ⌁

Vesica piscis is a Latin term which literally means fish bladder, referring to the lens shape formed by two circles which share each other's centre. A *vesica piscis* is a remarkable pattern; it is made up of two circles that overlap by the radius of each other, yet the *vesica piscis* hides within it perfect polygons and profound proportions.

To draw a *vesica piscis*, simply take a pair of compasses and draw a circle. Keep the compasses set on the same radius, place the point on the edge of your circle and draw another. You can add a construction line passing through the centres of your circles and out to the borders.

The *vesica piscis* is also referred to as a mandorla. Mandorla means almond-shaped, and it is a powerful form in religious art. There are numerous examples of the mandorla used to surround icons of Jesus Christ and the Holy Family in Christian art, representing heavenly glory, majesty and mystery. There is a sacred spring in Glastonbury, England, linked to the Grail Legend, which emerges in the Chalice Well gardens. Here the mandorla is celebrated on the well cover and in the overlapping pools where the healing water collects before it flows beyond the garden.

From Chartres Cathedral in France to Glastonbury Abbey in England to the central structure of the Angkor Wat temple complex in Cambodia, the circles of the *vesica piscis* seem to have been significant in setting out the plans of a great many sacred buildings. The overlapping circles suggesting the union of heaven and earth seems to have offered the correct proportions, while the central mandorla is often the focal point of power.

The *vesica piscis* represents union. The two circles are opposing energies brought into creative alignment: earth and sky, body and soul, physical and spiritual, masculine and feminine. The union is energized in the central overlap, suggesting the birth canal from which all life emerges.

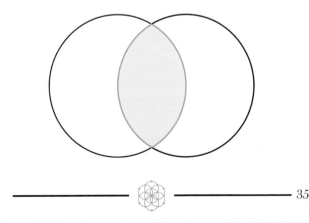

☉ INSIDE THE MANDORLA ☉

The *vesica piscis* or mandorla contains all potential of structure and form. That central region is where all geometry takes shape, just as young stars take shape in the swirling potential of distant nebulae. Let's begin by finding some triangles.

Draw or print out a picture of a *vesica piscis* and use a ruler to draw a line to connect the centres of the circles. From here it's easy to see how two equilateral triangles may be formed. Now extend the two upper sides of the topmost triangle downward until they meet the circumference of each circle.

Not only have you created four uniform equilateral triangles from just two circles, but together they form a fifth triangle which is also a net for a tetrahedron, an important four-sided triangular solid (see Platonic solids later in this chapter).

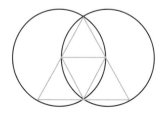

Discovering a further polygon from the triangles of the *vesica piscis* requires a third circle. From a standard two-circle *vesica piscis*, draw a line to connect the circles out to the circumference of one of the circles. This will give you a point for the centre of the third circle. With three circles there will be two overlapping *vesica piscis*. All you have to do is connect the points within the central circle to find a regular hexagon; itself formed of six equilateral triangles. From here, simply extending the hexagon edges to the circumferences of the outer circles will offer points to create a rectangle, which may be evenly divided into three. All polygons are waiting to be born in the *vesica piscis*.

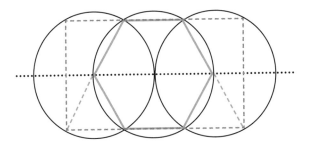

༄ **THE TRIANGLE** ༄

The triangle is the foundational polygon for all geometry. All regular polygons contain, or are composed of, the triangle. The five Platonic solids, which Plato understood to be the building blocks of all creation, present the triangle on all their faces. This is true even of the cube with its square faces since the square is composed of two or more triangles.

The triangle is the most rigid of geometric forms because each point on a triangle supports the other two equally. This is why cranes, pylons and many tall structures find their strength in triangles. The tensions, forces and energies of triangles are always in harmony, suggesting strength and support.

The triangle is often used to represent the Holy Trinity. A medieval motif used to explain the Holy Trinity placed God in the centre of a triangle to say that each person of the Trinity is God, while they are not each other. In this way, the triangle is a powerful symbol of supportive union. Two who are in a loving relationship always manifest a third energy, which is their strength and bond.

⌒ A UNION OF TRIANGLES ⌒

The hexagram is an extremely potent symbol used in a great many sacred contexts, from esoteric spirituality to Judaism. The hexagram is made up of two equilateral triangles – one pointing up and the other pointing down. To draw a hexagram, start with a circle, then overlap two others vertically so that the circumferences pass through the centre of the original circle. This will give six points to connect as two triangles.

The hexagram is seen as two polar opposites in union. The upward triangle may be recognized as the masculine, while the downward triangle is feminine. Similarly, the upward triangle is fire, while the downward is water. The upward is heaven, while the downward is earth. The hexagram therefore powerfully symbolizes the idea of "as above, so below".

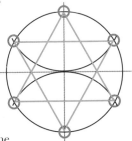

With the union of heaven and earth at its heart, the hexagram symbolizes a meeting of equals and a coming together of polar opposites.

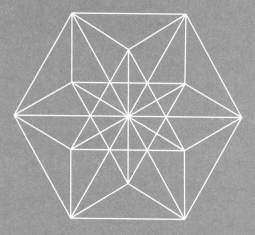

THE BODY SHOULD BE TRIANGULAR,
THE MIND CIRCULAR. THE TRIANGLE
REPRESENTS THE GENERATION OF ENERGY
AND IS THE MOST STABLE PHYSICAL
POSTURE. THE CIRCLE SYMBOLIZES
SERENITY AND PERFECTION, THE
SOURCE OF UNLIMITED TECHNIQUES.

MORIHEI UESHIBA

↶ THE SQUARE ↷

The square, being also a rectangle, is a basic shape for geometry. Its four points come to represent four seasons; the four elements of earth, wind, fire and water; and the four cardinal directions – north, south, east and west. With these associations, it is easy to see how the square represents the material and manifest world. It symbolizes stability, reliability, clarity and peace of mind.

There is something of a container energy to the square which makes it receptive. However, it can be a restrictive energy too as it acts as a boundary. This reflects our experience as human beings in many ways. Having a physical body, we are receptive to life through sensation, but we are constrained by the limits of physical form.

In the Bible, St John has a vision of a New Jerusalem, the mythical future city of God, which is built in a square. In fact, it is a perfect cube measuring 1,500 miles in length, width and height. St John writes that he witnessed an angel measure the city with a golden rod.

⌒ THE METATRON CUBE ⌒

With thirteen circles and an array of connected straight lines, the Metatron cube contains all geometric shapes within its matrix. This deeply intertwined matrix is named after an angel.

The Metatron angel of Judaeo-Christian tradition is the manifestation of the voice of God. The Metatron is an angel who carries the potencies of the divine utterance, the very creation of all things. In order to bring things into being, according to the book of Genesis, God spoke. When God finished creation, God said it was good, then created mankind in the divine image. The Metatron cube is therefore a symbol of the bridge between heaven and earth, the spiritual and the physical. Representing the divine voice spoken into all life, this symbol traces the sacred patterns encoded into all things.

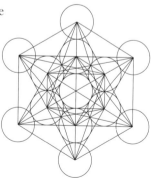

[THE UNIVERSE] IS WRITTEN
IN MATHEMATICAL LANGUAGE,
AND THE LETTERS ARE
TRIANGLES, CIRCLES AND OTHER
GEOMETRICAL FIGURES.

GALILEO GALILEI

⌒ **THE SEED OF LIFE** ⌒

Circles seem most fruitful when they intersect – when they live within each other while preserving themselves. The idea of fruitful circles is powerfully represented within one of sacred geometry's most important and iconic structures: the Seed of Life. The Seed of Life consists of six circles arranged around one.

To draw a Seed of Life, begin with a circle, then place the compass point on its circumference and draw another so that it passes through the centre of the first circle. Where circumferences cross, draw the next circle, and then the next until you have drawn all seven circles.

The Seed of Life is a powerful symbol of potential growth. The seed, with its seven aspects, has all that

is needed to flower. This fertile symbol invites us to conceive fresh ideas, new feelings, or even new life. It is a reminder that each of us has all we need to flourish.

The number seven is a significant number when it comes to creative energies. According to Genesis, the heavens and the earth were created in six days, then God rested on the seventh – represented by the six circles which surround one in the Seed of Life. When God sent the flood from which life was re-created, the sign of new promise was a rainbow made up of seven colours. Each of those colours is mapped to the seven chakras. Chakras are centres of life energy that run down the middle line of the body. Energy healing practitioners often work with chakras to promote spiritual and physical well-being. Turn to page 106 for a chakra healing meditation using sacred geometry.

With its seven circles corresponding to the seven chakras, the Seed of Life becomes a potent tool for unlocking the potential within us. Seeds are potential made manifest, and each of us carries the seeds of our creative potential.

⌒ **THE FLOWER OF LIFE** ⌒

The Flower of Life has the Seed of Life at its heart from which further circles radiate out, forming the petals of the inner border. The Flower of Life is a powerful mandala which invites meditation on the patterns of growth and the connectedness of all things.

There are 19 overlapping circles in all, symbolizing a kind of mutual and multiple interdependence. The spiritual teaching of the flower is that growth entails an awareness of dependency and respect. All living things have their place in an intricate network of symbiosis. Living in harmony with all life is to thrive as part of a beautiful whole.

Aristotle wrote that the goal of life was *eudaimonia*, which is best translated as flourishing or flowering. Flourishing requires the practice of balance and harmony in relation to others. It is about becoming your best and truest self.

❍ THE TREE OF LIFE ❍

Inside the Flower of Life, it is possible to discern another powerful symbol of sacred geometry: the Tree of Life. By connecting up the flower centres in the central region of the Flower of Life, a remarkably well-proportioned kabbalistic symbol emerges. This is the Tree of Life of Kabbalah – a school of Jewish mysticism.

The Tree of Life is made up of ten lights and 22 paths. The ten lights each enumerate a divine name, an essence of the otherwise unknowable God. These include wisdom, understanding, strength, love, beauty and kingdom. The 22 paths connecting them are interior gates through which to contemplate the divine. To learn more about God, we might contemplate a path connecting wisdom to understanding, for instance, or love to beauty. Some associate these paths with the 22 cards of the tarot trumps.

In the context of the Flower of Life, it represents the divine, hidden at the heart of all creative life.

⌒ THE SPIRAL ⌒

The spiral is that universal energy of growth, renewal and expansion. Organic growth seems to depend upon the spiral to make the best use of energy for strength and protection. From the spider web to the rose, spirals are an essential form.

Where the spiral represents expansive growth, it suggests limitless growth – so its source must be limitless divine abundance. It is a nurturing energy and a rooted one. The spiral grows out while remaining connected to its inner core.

In spiritual terms, the spiral offers reflection on the growth of the self in the same expansive way, fulfilling potential while continually renewing the source. The attainment of the higher self is a lifelong journey outward and upward while realizing the eternally true self that will always be found within.

The Wizard of Oz is a classic tale of that journey towards true self while discovering the self within. Dorothy is swept to her own dream lands by a tornado, itself a spiral, to find herself at the beginning of the yellow brick road – another spiral!

⌒ THE PLATONIC SOLIDS ⌒

Now that we have encountered the essential forms of sacred geometry in two dimensions, it's time to move to three dimensions. The basic shapes in 3D are the five Platonic solids – these were assigned eternal and creative importance by the philosopher Plato. We'll learn more about him in the next chapter.

The Platonic solids are formed from the regular polygons of triangle, square and pentagon. The hexagon is not included because it's impossible to create a solid from a net made up of hexagons. There are, and only can be, five essential solids. Each solid depends on a net of regular polygons, and only five of these will close completely into one regular three-dimensional form. Sounds unlikely, but it's true!

The Platonic solids are the building blocks of creation. These sacred shapes discovered by ancient mathematicians are symbols of eternal truth. They unlock a reality within us when they are realized, which is to say, made real. With the Platonic solids, these energies take a three-dimensional shape, which makes them part of our lived reality.

⌒ **THE TETRAHEDRON** ⌒

The tetrahedron is made up of four equilateral triangles. It is associated with the element of fire and the colour yellow. Spiritually speaking, tetrahedron energy is all about manifestation.

No matter how a tetrahedron is rotated, it always points to the sky with a firm base on the ground. This is why its vibration resonates with manifestation and awakening. The energies of creative emanation are brought to be with a solid foundation. The tetrahedron is therefore the perfect vibrational energy to draw upon if you want to bring about some changes in your life. It will harmonize with your personal power, support acceptance of the past and invite openness to fresh opportunities. Quartz crystals, known to be crystals of spiritual awakening, grow in a tetrahedron structure.

◌ THE HEXAHEDRON (CUBE) ◌

The hexahedron is made up of six squares. It is associated with the element of earth and the colour red. In spiritual terms, the hexahedron energy is all about grounding.

The hexahedron always rests on a firm, square foundation. This is why its vibration resonates with safety, security, and survival. These energies invite us to reconnect with the earth, with nature, and with the spirit of Gaia. This is an energy to turn to when we are feeling anxious, afraid, stressed or just a bit lost. This is the earth and all its cardinal compass points, helping to centre us and allow us to see all paths. This is also a foundation energy – a solid place to start from. In this sense, the hexahedron offers confidence to spark creativity.

Pyrite crystals grow in a hexahedron structure. Pyrite is a protective stone that grounds and supports.

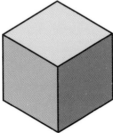

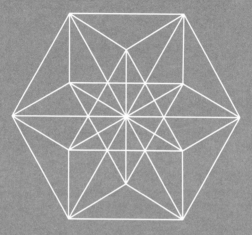

LEARN HOW TO SEE. REALIZE
THAT EVERYTHING CONNECTS
TO EVERYTHING ELSE.

LEONARDO DA VINCI

↷ **THE OCTAHEDRON** ↶

The octahedron is made up of eight equilateral triangles. It is associated with the element of air and the colour green. Spiritually speaking, the energy of the octahedron is all about integration.

The octahedron is formed of two square-based pyramids brought together in a way that binds yet preserves and strengthens separate identities. This is why its vibration resonates with togetherness, matters of the heart, our feelings, emotions and love. These energies invite us to open the heart, to seek togetherness, and to reach out to one another. The octahedron is an energy to draw on when uncertain about a relationship. This might require forgiveness for healing or self-sacrifice and care to deepen love. The integration energy also applies to the self, since finding love and respect for yourself is an essential part of being whole.

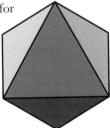

Fluorite crystals grow in an octahedron structure. Fluorite integrates spiritual energies and opens the heart.

⟳ **THE ICOSAHEDRON** ⟲

The icosahedron is made up of 20 equilateral triangles. It is associated with the element of water and the colour orange. In spiritual terms, the energy of the icosahedron is all about transformation.

The icosahedron is the closest of the Platonic solids to a sphere and represents the halfway point on a journey of self-transformation. This is why its vibration resonates with self-realization, self-actualizing, individuation and the journey of becoming. These energies invite us to elevate the self towards a new phase in consciousness. The icosahedron therefore unlocks potential and helps remove blocks to spiritual fulfilment.

There is something of the alchemist at work with this energy. It is an invitation to begin transforming yourself to fulfil your highest destiny.

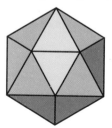

Amethyst geodes reveal points which are triangular-faced like the icosahedron. Amethyst is a crystal which is supportive during periods of change.

↷ THE DODECAHEDRON ↶

The dodecahedron is made up of 12 pentagons. It is associated with universal matter or ether, and the colours violet, indigo and purple. Spiritually speaking, the energy of the dodecahedron is all about ascension.

The dodecahedron is the glitter of the heavens – the stars that make up the 12 signs of the zodiac. The dodecahedron's vibration resonates with the fifth element: the quintessence or soul. It could be thought of as the substance of soul or the stuff that souls are made of – not just our individual souls, but the soul of the universe itself. This energy invites us to ascend, to transcend the limitations of our human bodies and connect with a higher self. The dodecahedron is therefore an invitation to meditation, to detach from ego and find union with the divine.

Garnet crystals grow with a dodecahedron structure. Garnet cleanses and re-energizes mind and body during meditation.

☾ THE MERKABAH ☾

While it is not one of the Platonic solids, the Merkabah is a three-dimensional shape made up of two tetrahedrons. It is also known as the star tetrahedron. The Merkabah name comes from ancient Jewish mysticism based on a vision of God seen by the prophet Ezekiel in the Bible. Ezekiel is said to have seen a divine chariot, or Merkabah, that carried people to heaven. Scholars of Jewish mysticism have come to believe that the chariot was shaped as a star tetrahedron. It is intriguing to notice that, seen from the side and represented in two dimensions, the Merkabah looks like a hexagram or Star of David.

The Merkabah may be thought of as a light body which all of us have. It is our own chariot or transport between worlds. The first step of Merkabah awareness is to activate your Merkabah. This can be done through meditation, visualization or simple, intuitive commands such as "Dear Merkabah, please heal me."

A BRIEF HISTORY
OF SACRED
GEOMETRY

On the banks of the Nile, 4,500 years ago, craftsmen placed the capstone on the Great Pyramid, completing four enormous triangles to form a sacred tomb. Approximately 100 years later, on Salisbury Plain in England, massive rectangles of stone marked out the sacred circle of Stonehenge. Another 4,000 years later, in the seventeenth century, Mughal emperor Shah Jahan looked over the plans for his wife's sacred mausoleum and saw a square upon a square: the Taj Mahal. Plain geometry, simple symmetry and divine proportions have outlined sacred space for as long as the ideas of triangle, circle and square have persisted in the mind.

We have already met one important figure in the long history of sacred geometry, Leonardo Bonacci, who came to be known as Fibonacci. His writing brought the Fibonacci sequence to European attention in 1202. The Golden Ratio is even older. An ancient Greek mathematician named Euclid wrote about that some 2,300 years ago. In this chapter, we will meet more key figures who have contributed to this fascinating history.

LET NO ONE IGNORANT OF GEOMETRY ENTER HERE.

A SIGN RUMOURED TO BE OVER THE DOOR OF PLATO'S ACADEMY

A UNIVERSE MADE FROM NUMBERS

We begin our story of sacred geometry around 500 BCE with Pythagoras. Pythagoras is probably one of the best-known Ancient Greek philosophers because of his famous theorem concerning triangles. Many of us can remember learning that for a right-angled triangle, the square on the hypotenuse is equal to the sum of the squares on the other two sides. Yet Pythagoras also held some compelling philosophical beliefs concerning the order and structure of the universe.

For Pythagoras, number was the underlying substance of reality, so mathematics and geometry held the key to discovering the true nature of the universe. Pythagoras' point was that the universe contains a great many things, and the one thing they all have in common is that there are a number of them, and they can be counted. Everything in the universe can be counted: from the grains of sand on the shore to the hairs on a person's head. Even the universe itself can be counted; there is only one (unless you subscribe to multiverse theories). For Pythagoras, this meant number was fundamental to the universe.

THE PYTHAGOREANS AND THE HOLY TETRACTYS

The followers of Pythagoras were so enchanted by number, pattern, and shape that they directed prayer and ritual through and to numeric structures. The most significant numeric structure for the Pythagoreans was the holy Tetractys. The Tetractys is an arrangement of four rows of numbers shown as dots to represent the numbers one, two, three and four:

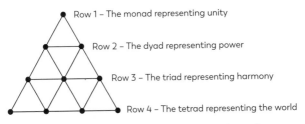

Row 1 – The monad representing unity

Row 2 – The dyad representing power

Row 3 – The triad representing harmony

Row 4 – The tetrad representing the world

The Tetractys represents all of creation through every geometric dimension. The monad is the point, the dyad is the line, the triad is the surface, and the tetrad is the solid. Most importantly, the Tetractys contains the decad or ten, the holiest of numbers, because the sum of the Tetractys numbers, one, two, three and four, is ten. For the Pythagoreans, the Tetractys was a key to the underlying divine principle of the universe, which is why it was worthy of prayer and contemplation.

THE INFLUENCE OF THE HOLY TETRACTYS

The Tetractys is an ancient numeric structure, and its influence in esoteric spirituality is considerable. Why is this structure so complete? All the numbers one to nine required to reach the perfect number ten are found within the Tetractys: 1, 2, 3, 4, then 4+1=5, 4+2=6, 4+3=7, 4+3+1=8, 4+3+2= 9 and 1+2+3+4=10.

In the Tetractys, there are four things of a similar nature – four numbers, four rows – and together they make a fifth, which is their totality. That number five again! This accords with four material elements, fire, water, air, earth, and the fifth, which is spirit. Again, the same is true of the Platonic solids – four material solids and the dodecahedron, which is the substance of the universe. In a way, the dodecahedron *is* the universe that the material solids exist within.

It is fascinating to find that the tarot follows a similar structure – twice! There are four suits of the minor arcana – wands, cups, swords and pentacles – and then there is the major arcana. There are four court cards in each suit – and then ten number cards.

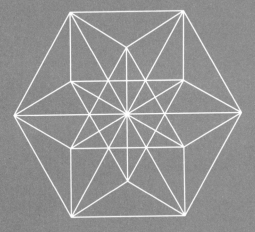

BLESS US, DIVINE NUMBER,
THOU WHO GENERATED GODS AND
MEN! O HOLY, HOLY TETRACTYS,
THOU THAT CONTAINEST THE
ROOT AND SOURCE OF THE
ETERNALLY FLOWING CREATION!

A PYTHAGOREAN PRAYER

THE PARTHENON – DIVINE ARCHITECTURE?

Probably the most famous building of Ancient Greece is the Parthenon. Completed in 438 BCE, it was built as a temple to the goddess Athena. Just as Stonehenge honours the circle, and the Great Pyramid of Giza honours the triangle, so the Parthenon honours the rectangle.

The most intriguing thing about the design, however, is that it has no straight lines and no right angles. Of course, looking at the Parthenon, it certainly seems to follow a rectangular design. The architects realized that parallel lines appear to bow when seen from a distance, especially when looking up at them. Correcting this optical illusion meant not using straight lines or right angles. Only in this way could the Parthenon truly honour the rectangle.

At the time of the construction of the Parthenon, the principles of beauty in structure and form were summed up by the sculptor Polykleitos as harmony, symmetry and proportion. These principles applied for human forms in marble as well as grand temples.

With proportion, symmetry and harmony in mind, we might expect it to be golden in ratio and rectangle. Many observers certainly think that it was designed that way, while others feel any golden ratios are coincidence. In the picture below, the façade of the Parthenon is shown to be a golden rectangle with an inner series of rectangles following the familiar logarithmic golden spiral. Of course, if the roof was intact, the height proportion would complete the picture and confirm a golden rectangle.

PLATO, IDEAL GEOMETRY AND THE PLATONIC SOLIDS

Our next philosopher is Plato, from around 400 BCE. Plato, like Pythagoras, thought of reality as divided into the intelligible and the sensible realms. The sensible realm is the reality we encounter through our five senses. The intelligible realm is an eternal realm that philosophical reasoning leads us to.

Plato reasoned that creation consists of imperfect copies of a set of perfect ideas or forms. These forms are eternal, very real, and exist in the intelligible realm like templates in the mind of God. Some of these forms are high ideals such as goodness or beauty. Others are geometric: triangles, circles and squares. Plato believed that when the demiurge (Plato's creator demi-God) created the world, he worked with these perfect forms as best he could. Good things in the sensible realm can only point to perfect goodness. Circular things in our world are never quite the perfect circle.

This means that perfect geometric shapes belong in the eternal realm, alongside the forms of goodness and beauty. It also means that in practising geometry we get to glimpse the divine templates of creation.

In his *Timaeus* dialogue, Plato begins with an elemental triangle, a right-angled triangle with the other two angles at 45°. Adjoining two of these on a straight edge gives an equilateral triangle, which Plato finds to be the fairest of all. Plato shows how this equilateral triangle may form a tetrahedron. This is the first of the Platonic solids. Eight of the same triangles create the octahedron.

Then Plato describes a solid with 20 faces, made up of 20 tetrahedrons: the icosahedron. The isosceles triangle is shown to make up the cube, the fourth of the Platonic solids. Plato believed the demiurge used these first four solids as the building blocks of creation – the elemental molecules. The tetrahedron is the molecule of fire, the octahedron is the molecule of air, the icosahedron is the molecule of water, and the cube is the molecule of earth.

The fifth of these solids is also made up of triangles, having 12 pentagonal faces: the dodecahedron. Plato describes how the demiurge had this one left over, so he used it to decorate the heavens.

◯ **THE HOLY MANDALA** ◯

While ancient Greek philosophy and architecture were discovering new realities in number, proportion and geometric form, ancient Hindu and Buddhist consciousness was drawn towards the spiritual power of geometric pattern. Mandalas have been part of meditative practice in Asian culture for over 2,000 years.

A mandala is a powerful spiritual and ritual symbol. Mandalas are commonly circular, with geometric patterns forming polygons and other shapes derived from the circle. Indeed, the word mandala is Sanskrit for circle.

Mandalas may be seen as a visual representation of the universe, or as a gateway to the inner self in meditation. By meditating with a mandala, we are drawn to its centre. In doing so, we participate in a transformative journey, one which changes the self and leads to a more profound awareness of our place in the universe.

The mandala pictured here is the Sri Yantra or king mandala. It is considered the most important of geometric symbols, representing the sacred sound of creation – the Om.

☙ ISLAMIC GEOMETRY ❧

Sacred geometry has been an essential part of the structure and decoration of Islamic holy places since at least the eighth century. Muslims do not allow representations of the human form in religious art: they believe the creation of the human form is for Allah alone. For this reason, Muslim tradition has been to decorate the interiors of mosques, towers and palaces with colourful geometric patterns in mosaic.

The underlying form of the circle provides the structure for Islamic mosaics. This in turn provides the square, triangle and multisided polygons which seem to repeat forever. From these emerge arabesque designs with intertwined flowing lines, rhythmic linear patterns, and scrolling interlaced foliage. Arabesque patterns depict growth in nature, while the underlying circle represents primordial unity from which all creation flows.

It is easy to lose yourself while gazing at the beautiful patterns of Islamic mosaics. In losing the self, a Muslim may be inspired to experience oneness, which is the essence of Allah. A saying of the Prophet Muhammad shows how beauty is considered divine: "God has inscribed beauty upon all things."

CHARTRES CATHEDRAL –
SACRED GEOMETRIC SPACE

Chartres Cathedral was built in the eleventh century by master masons. It is a sacred space formed on the principle of sacred geometric design to the tiniest detail of divine significance. The whole plan of the building fits perfectly within a *vesica piscis*, as does the great Belle Verrière window, which depicts the Madonna and Child.

The Cathedral's other famous window is the North Rose Window, itself a magnificent example of sacred geometric design. The window is structured around the number 12. 12 is a very significant number spiritually and geometrically. Remember the 12-sided polyhedron which Plato said God used to decorate the heavens? There are 12 signs of the Zodiac which map the night sky. There are 12 tribes of Israel, according to Jewish tradition. Jesus chose 12 apostles, and there were 12 baskets of food left over when Jesus miraculously fed the 5,000.

The design of the window emerges from some simple work with a circle. To give this a try, you'll need a compass. First, draw a circle, then find the four quarter points and, using the same radius, draw four curves inside. This will mark out 12 points on the circumference.

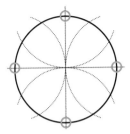

Remember that magic number five in sacred geometry? Five Platonic solids, five points on the pentagram, five senses and five elements? Draw a straight line between every fifth point on the circumference to create a dodecagram.

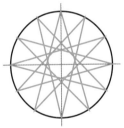

Our dodecagram gives the basic geometric structure for the North Rose Window of Chartres Cathedral. It also makes a wonderful geometric shape with sacred numbers encoded within it. Take a look at the North Rose Window online for inspiration to colour and complete your diagram.

LEONARDO DA VINCI
AND THE RENAISSANCE

We now come to the turn of the fifteenth century, and Leonardo da Vinci, one of the most widely celebrated and influential figures of the Italian Renaissance. His work leads into a significant period in art history known as the High Renaissance with *The Last Supper* painted at the beginning of this period and his most famous work, the *Mona Lisa*, painted at its height.

Among the defining features of High Renaissance works of art is a kind of physical realism, which is achieved through the careful use of perspective and balanced composition. Perspective offers pictures realistic depth, while balanced composition of all the essential elements of a picture gives a harmonious sense of proportion. Those who have studied Leonardo's art have found his compositions often depend on the Golden Ratio, which lies at the heart of sacred geometry. Golden rectangles and golden triangles seem to be everywhere in Leonardo's work. Even the proportions of *Mona Lisa* herself seem to follow a Fibonacci spiral (or logarithmic golden spiral) – as long as you start with her right nostril!

LEONARDO, PACIOLI, AND THE DIVINE PROPORTION

For Leonardo da Vinci the Golden Ratio went beyond a balanced-looking rectangle or a well-adjusted triangle. It was encoded into all the most elegant structures. It formed the basis of the most stable and striking architecture, but it also underpinned the proportions of the human body and mapped the human face. Accordingly, the closer the proportions of the face agree with the Golden Ratio, the more beautiful a person appears.

The idea that the form of the human body depends on a specific proportion inspired the minds of Renaissance artists toward the great architect – God. In 1509, Italian mathematician Luca Pacioli published a work entitled *The Divine Proportion*, for which Leonardo provided illustrations of the Platonic solids.

Pacioli's book is a detailed study of the Golden Ratio in geometry, architecture, and the structures of the human body. The book also includes reasons why the Golden Ratio should be named the divine proportion. The divine proportion is the proportion of three lengths, suggesting the Holy Trinity. It is an irrational number, suggesting divine incomprehensibility, and it is a single universal value, suggesting divine simplicity.

HOW TO MAKE A PLATONIC SOLID USING DIVINE PROPORTIONS

In his book *The Divine Proportion*, Pacioli describes how to make one of the Platonic solids drawn by Leonardo using three interlocked golden rectangles. This is a fascinating way to see how perfect proportions support the construction of sacred shapes. The icosahedron is a 20-sided shape, one of the five solids identified by Plato.

Pacioli's description goes something like this:

1. Take two golden rectangles of the same dimensions and place them at right angles to each other. Push one inside the other so that their centres meet:

2. Take a third golden rectangle at right angles to the first two and push this into the other two so that their centres meet:

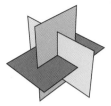

3. To make the 20 triangle faces of the icosahedron, simply join up the vertices (points) in threes:

Pacioli illustrates perfectly how divine proportion lies at the heart of the Platonic solids. Beginning with three golden rectangles, it is possible to find 20 perfect triangles, arranged in the most efficient way possible from just 12 shared vertices, creating a perfectly proportioned icosahedron. Sometimes, it really is like the universe is trying to tell us something!

PACIOLI, LEONARDO AND VITRUVIUS

Pacioli and Leonardo followed in the footsteps of a Roman architect from the first century BCE named Vitruvius. Vitruvius believed that buildings should be built around three key principles: strength, utility and beauty. Temples particularly should be designed with carefully balanced symmetry and proportion: the parts should correspond agreeably with the whole. In one important sense, the body is a temple for the soul – a temple with a divine architect – so it made sense that temples for the divine should follow the proportions of the human body.

Vitruvius took measurements of the human body to find ideal proportions and symmetry. He then used these as scale measurements for temples. He noticed particularly how the stretched-out body of a man might be circumscribed by both a circle and a square. In an X-shape, the circumference of the circle, with the navel as its centre, is touched with fingers and toes. In a T-shape, the perimeter of a square centred on the navel is touched with head, feet and fingers. These observations fascinated Leonardo and led to his famous picture: *Vitruvian Man*.

BEAUTY IS PRODUCED BY THE
PLEASING APPEARANCE AND
GOOD TASTE OF THE WHOLE,
AND BY THE DIMENSIONS OF
ALL THE PARTS BEING DULY
PROPORTIONED TO EACH OTHER.

VITRUVIUS

LEONARDO DA VINCI'S ART AND THE DIVINE PROPORTION

Take a look at some of Leonardo's artwork to see how he used the divine proportion to give his pictures perfect composition. One famous work to search out online is *Annunciation*, painted around 1472.

From left to right, notice how the corner of the brickwork is almost exactly 62 per cent of the full width of the picture. The divine proportion becomes easier to spot when you think of it as just under two-thirds. In this case, the corner of the building is just under two-thirds of the picture width. Now look at the picture from right to left. The edge of the lower wall to the garden and beyond is almost exactly 62 per cent of the full width of the picture.

This isn't just about a pretty picture – it has real sacred significance. The two people in the picture are divine: the Angel Gabriel on the left and Mary, Mother of God, on the right. Both representations of the divine fit perfectly with the divine proportion. From the left, Mary is two-thirds of the picture width; from the right, Gabriel is. Leonardo's *Vitruvian Man* is in divine proportion in a fully geometric sense. The height of the figure forms a square with the outstretched span of the arms. With outstretched legs and upward-stretched arms, the figure touches the points on a circle with its centre at the navel. There is a circle and a square outlined in Leonardo's picture, but no triangle as such.

Leonardo writes that an equilateral triangle will be found with its apex at the navel tracing the outstretched legs. Yet there is another intriguing triangle to be discovered. Draw a circle within the square, then draw an equilateral triangle that touches this circle at three points, with the apex at the top of the man's head. This triangle shows how the eyes, collarbone and pectoral muscles relate to each other in a golden ratio proportion. The sides of this implied triangle run in parallel to the triangle of the outstretched legs indicating further symmetry and proportion for the whole body.

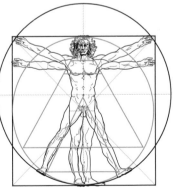

JOHANNES KEPLER AND THE PLANETS

Seventeenth-century astronomer Johannes Kepler has been described as the last great Pythagorean. In *Mysterium Cosmographicum*, Kepler wanted to show how he believed God used the five Platonic solids to construct the universe.

Kepler explored the relationships between the planets and their distances from each other. He found that their relative distances corresponded with concentric spheres nested within each other, separated by each of the five Platonic solids. To visualize this, think of the solids as Russian dolls each wrapped in their own closely fitting sphere. With the sun in the centre encased within an octahedron, inside an icosahedron, inside a dodecahedron, then tetrahedron, then cube, six layers are produced. These layers represent the relative distances between the six planets known to Kepler – Mercury, Venus, Earth, Mars, Jupiter and Saturn.

Even though the theory turned out to be approximate, Kepler's work did reveal three important laws of planetary motion based on his discovery that planetary orbits are elliptical. Most significant for Kepler's astronomy is the way he viewed the universe as harmonious and ordered, built by geometric design.

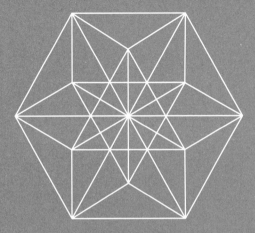

WHAT IS GOD? LENGTH, BREADTH,
HEIGHT, AND DEPTH.

ST BERNARD OF CLAIRVAUX

KEPLER, THE SNOWFLAKE AND THE BEE

In 1611, Kepler wrote an essay entitled *On the Six-Cornered Snowflake*. Kepler couldn't decide what to buy as a Christmas gift for a friend. Noticing a snowflake on his coat, he wondered at its six-cornered symmetry: why should a snowflake fall with six corners and not five, or seven? The beauty of the snowflake inspired him to reflect on a range of structures considered beautiful in nature. He considered flowers, the Fibonacci sequence and growth patterns. He examined the arrangement of seeds in a pomegranate, the geometry of tessellation, and the architecture of beehives.

The cells in which bees live are hexagons, which tessellate perfectly to make best use of space. More interesting to Kepler were the rear walls of the cells, which are not flat. The rear walls are formed from geometric shapes known as rhombi, which are connected together. This enables the bee's cell to adjoin ten neighbours: six around the sides and four at the rear.

Kepler realized that bee cell structures maximize shared walls – they are as efficient as possible to construct, and bees can work together effectively to create them. Kepler took his observations about bee cell architecture and devised a new solid figure constructed of 12 rhombi, which, just like the cube, can fill three-dimensional space without any gaps. This is known as Kepler's rhombic dodecahedron.

When Kepler examined the cosmos, he saw divine and geometric patterns everywhere. He saw the movement of the planets, the cities of bees, and the structure of a single snowflake as evidence that sacred geometry underpinned the whole of creation. And that Christmas present... What better than a snowflake? A little star fallen from heaven.

◠ AN INFINITY OF TRIANGLES ◠

In the first half of the twentieth century, Polish mathematician Wacław Sierpiński worked on fractals (geometric shapes with repeating properties), revealing fascinating insights into infinity and geometry. The fractal design known as the Sierpiński triangle perfectly illustrates the way in which triangles create triangles just like an infinity mirror – triangles inside triangles disappearing into forever.

To draw a Sierpiński triangle, begin by drawing an equilateral triangle. Connect halfway points on the perimeter to form an inner equilateral triangle. This divides the original triangle into four. Leaving the central triangle in place, subdivide the surrounding triangles in the same way, then simply keep going.

The Sierpiński triangle is a powerful symbol of the creative potential of triangles and the incredible way in which geometric shapes seem ready to burst into life. The Sierpiński triangle can be automatically generated through a very simple algorithm or chaos game. It's worth searching online for an animation of this algorithm.

⌒ **SACRED GEOMETRY TODAY** ⌒

In the twenty-first century, awareness of sacred geometry has grown considerably. More and more people are coming to a new kind of divine consciousness, and sacred geometry helps us develop and experience this.

Sacred geometry healing is practised today in many different forms. Some of the more elaborate techniques involve the use of a full-size Merkabah (see page 56). During a healing session, the individual steps into the Merkabah and is guided through meditation and sound frequencies towards an elevated state of consciousness. This can help with spiritual growth or for specific healing, such as releasing negative emotions.

As we entered the twenty-first century, crop circle researchers noticed growing complexity in designs – these following sacred geometric principles. Crop circles continue to appear yearly near ancient sacred sites in the UK.

With consciousness emerging, many people are finding sacred geometry a helpful way to remain connected to divine abundance. Sacred geometry jewellery and tattoos can help some people raise their consciousness and resonate more deeply with life.

PRACTISING
SACRED
GEOMETRY

Having encountered the sacred shapes and learned how our ancestors were profoundly influenced by sacred geometry, we can now explore some ways to invite the power of sacred geometry into our lives. In this chapter you will find meditations to help you gain deeper insight into the sacred shapes, to align yourself spiritually with their resonance and to access their power for healing.

Sacred geometry is all about sensing connectedness and allowing the sacred shapes to raise us to a higher consciousness. Enjoy a more profound relationship with sacred geometry as you work through these activities.

TRY SOME SIMPLE CRYSTAL GRIDS

The shapes of sacred geometry can be experienced energetically through the use of crystals. Making a crystal grid doesn't need to be complicated. To begin with, try some simple crystal arrangements in a circle, triangle or square.

Start with the sacred shape you want to work with, drawn or printed onto card. This helps focus the mind and align the crystals most effectively.

- Choose a circle grid to meditate on oneness, self-renewal and regeneration.

- Choose a *vesica piscis* grid to meditate on union, creative potential and abundance.

- Choose a triangle grid to meditate on harmony, strength and support.

- Choose a square grid to meditate on stability, clarity and peace of mind.

- Choose a spiral grid to meditate on growth, life's pilgrimage and the path to the higher self.

- Choose a pentagram grid to meditate on balance, manifestation, discipline and the empowerment of the self.

- Choose a hexagram grid to meditate on the divine union, heaven in harmony with earth, "as above, so below."

Choose crystals that you feel will best support the energy you wish to draw on. Place your selected crystals on the points of your chosen shape or around its outline. Rounder stones will sit well on angles while crystal wand or point shapes can be placed on lines.

Position yourself comfortably in front of your crystal grid. Once you're relaxed, allow your gaze to rest gently on the grid. When you feel you want to, close your eyes and open your mind. Try not to think about this too hard; allow images and ideas to come to you naturally.

You can be as creative as you'd like with these grids. Meditate for as long as feels right, but expect to take at least ten minutes to enter a rested and receptive state. As soon as you have completed your meditation, write a journal entry as freely as you can. This can often help to crystalize any insights that may come to you. You could set up a grid somewhere in your home to invite the energy into your living space.

DEVELOP COMPLEX CRYSTAL GRIDS

More complex crystal grids make use of the more intricate sacred shapes and allow for a deeper power of intention intensified by creative visualization techniques. A powerful grid, which is also fairly straightforward to set up, is the Seed of Life grid.

The Seed of Life sacred shape allows for 13 crystals to be placed on its centre and intersections. To empower this grid most effectively, select one crystal for the central position, six of another type for the inner circle intersections, and six of a further type for the outer intersections. Arrange these on a drawing or print of the Seed of Life.

The central stone acts as the focus stone, gathering and drawing energy into the grid. The stones immediately surrounding the focus stone are known as way stones, which direct and modify the energy, radiating it through the grid. The stones on the outside perimeter of the grid are known as desire stones. They radiate the energy further in alignment with your intention or goal for this grid.

It's a good idea to cleanse the stones beforehand using your preferred method: running water, rock salt or by moonlight. Once you are ready to work with your grid, you might like to clear the space using some incense, such as sage.

Once you feel relaxed and comfortable, begin your practice by focusing your intention. You may wish to call on the healing energy of the Seed of Life through regeneration and renewal. Or you may wish to call on the fertility energy of the seed for new life or fresh ideas.

You can further empower grids by visualizing a stream of pure white light pouring into the focus stone. See the light trace the paths of the grid and flow into the way stones. As the light passes through the way stones, it changes colour and travels along all paths to the desire stones. See the light pulsating brightly in each of the desire stones, bringing your intention to fruition. Spend time in meditation, sensing and connecting with this energy.

TRY A PLATONIC SOLID MEDITATION

The Platonic solids are potent structures for meditative practice. A Platonic solid meditation can help you gain further insight into the vibrational energy of each solid. It can also help you connect with this energy more readily as you become more attuned and receptive to it.

The goal of a Platonic solid meditation is realization (in the sense of making real). Each solid is made real through creative visualization. When this is achieved, it provides a space to realize or make real the self in relation to this energy. All meditation is ultimately aimed at self-realization. In this sense, it raises consciousness of the self towards connection with your higher self. Platonic solid meditation offers an alternative pathway towards this raising of consciousness.

Each Platonic solid is associated with its own element, colour and teaching. Use the table on the opposite page to select the shape that best matches your intention. When you have done this, turn over the page for details on how to try platonic solid meditation.

Shape	Sides	Element	Colour	Teaching
Tetrahedron	4	Fire	Yellow	Manifestation
Hexahedron	6	Earth	Red	Grounding
Octahedron	8	Air	Green	Integration
Dodecahedron	12	Ether, Universe, Spirit	Violet, Indigo, Purple	Ascension
Icosahedron	20	Water	Orange	Transformation

First, we need to try to relax our minds and bodies. If you are new to meditative practice, begin by sitting in a comfortable chair in an upright position with your feet flat on the floor. Take several deep breaths all the way in and out while consciously relaxing each part of the body in turn, from head and shoulders down to your feet. Close your eyes when you feel ready.

Next, begin to visualize the chosen Platonic solid forming around you in its associated colour. Visualize its edges as bright lines of light in that colour. See yourself in the centre of the Platonic solid. Spend time feeling the presence of this form. Sense its energy as warmth radiating within you.

After a time, you can begin to notice the space, or the place, outside your Platonic solid. Allow yourself to explore. It might be that you step onto a path, through a doorway, or even float on an ocean. It may be a place you recognize, or it may be just a feeling. Relax and allow whatever comes to you to unfold. The energies of the Platonic solids are always positive and nurturing. Make a journal entry: what images or emotions came to mind? What impressions remain with you?

◠ DRAWING MANDALAS ◠

The twentieth-century Swiss psychoanalyst Carl Jung was fascinated by mandalas. As he began to draw them regularly, he found they revealed a great deal about the state of his inner self. In time he came to believe that the mandala acted as the centre for the self and that the self was always at the centre. In this way, he found mandalas provided pathways to the centre, to the self, and to the ultimate state of self-awareness and self-actualization, which he called individuation.

Try drawing a mandala each day. You could even take up the challenge to draw 100 mandalas over 100 days. Those who draw mandalas find it beneficial in a number of ways, including:

* For relaxation

* To improve focus and stimulate creativity

* To help with stress, anxiety, fear or depression

* To improve mood

* To increase peace and connection

* For self-acceptance

* For fun!

HOW TO DRAW A MANDALA

There are two approaches to drawing mandalas. You can draw them freehand, or you can use geometric tools to construct them. As a daily spiritual exercise, it's probably better to work freehand. This will enable you to feel a greater sense of freedom, which will encourage creativity and allow the subconscious to come through more naturally. Your mandalas don't need to be perfect!

Take a piece of paper and a pencil. (Good-quality art paper is best.) You're going to need some colouring pencils too. Watercolour pencils work well, as you can soften this with a paintbrush afterwards.

Start at the centre with some basic lines:

Once you have the basic central shape, it is simply a matter of outlining:

The alternative to outlining is interleaving:

Outlining and interleaving continue any way you like. And when you're done, you can add some colour:

Of course, as you grow in confidence with your drawing, you need not be restricted to outlining and interleaving. All sorts of patterns, from swirls and spirals to curves and labyrinths, emerge in the mandala, and the freer you feel to express yourself through these, the better. If you decide to draw a daily mandala, keep a brief journal entry alongside your mandala. Note your mood and any thoughts or ideas the mandala has unlocked for you. Has an issue become clearer for you? Have you released any negative emotions you might have been holding on to?

⌒ **KING MANDALA MEDITATION** ⌒

The Sri Yantra or king mandala (see page 69) is a potent sacred symbol. The word yantra refers to its power to bring freedom from bondage. This mandala therefore invites freedom from the aspects of self that restrain or constrict. These might be rooted in fear, cynicism or harsh self-criticism. The king yantra unlocks your true self from the various archetypes we construct and the personas we adopt in our interactions with the world outside. Spending time in meditation with the king mandala can lead us towards the point of certainty within.

To meditate with the king mandala, we first need to try to relax. Find a comfortable seating position with a picture of the king mandala in front of you. As always, begin your meditation with several deep breaths. Relax your whole body, from the top of your head down to your toes. If you notice any tension, try to let it go.

When you are ready, gaze at the mandala with relaxed eyes. There is no need to strain or stare – you can allow your eyes to blink as normal. Just let your gaze fall into the pattern. As you gaze at the symbol, allow your mind to be still. Release thoughts if they come to you and gently bring your attention back to the mandala. Focus on your breath, breathing steadily but naturally. After a time, you will become peacefully absorbed into the energy of the mandala.

Remain in this relaxed and receptive mode for as long as you wish. You should feel totally peaceful, perhaps with a sense of profound balance. As soon as you complete your meditation, make a journal entry. What images or insights come to you? Write freely to discover what the king mandala has unlocked for you.

◠ A GEOMETRIC TAROT SPREAD ◡

There are a number of ways to incorporate sacred geometry principles into tarot. This suggested tarot spread follows the Pythagorean Tetractys (see page 61) with its four rows of the first four numbers: one, two, three and four.

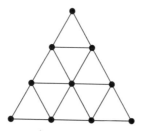

Hold the intention of the reading in mind (whether it's for yourself or on behalf of someone else), while laying out ten cards face down in four rows.

Read the cards by revealing one at a time and interpreting it in relation to the row it is in. This spread can be read in two ways depending on your intention. Read the cards from top to bottom to discover how a situation will unfold, or from bottom to top to get to the heart of an issue.

- The tetrad row: This is the manifest world or the square. This row concerns the setting of the issue – its situation. It will outline the key aspects of the matter and its primary concerns. It will also reveal how an issue, a desire or a plan has come to manifest. Draw four cards for this row.

- The triad row: This is the realm of form or the triangle. This row concerns the energies that persist over the situation. This might reveal the challenges, anxieties or obstacles. It might reveal the desires, motivations or inspirations. Draw three cards for this row.

- The dyad row: This is the realm of creativity or the *vesica piscis* (see pages 34–35). This row concerns the opposing forces which drive, initiate or underpin the issue. It might reveal forces that must be balanced and harmonized to resolve the issue or highlight energies that will offer the strongest support through the issue. Draw two cards for this row.

- The monad row: This is the ultimate concern or the circle. This row concerns what lies at the heart of the matter. It might reveal the ultimate outcome of the issue or its source. Draw one card for this row.

◠ MAKE A TETRACTYS TALISMAN ◠

A Tetractys talisman is a talisman of manifestation. It will help you manifest your highest intention for spiritual growth, healing, protection or support through difficult times. It can also align you with divine abundance to realize wealth or open up opportunities for you. Prepare some suitable card, leather, fabric, parchment or even a stone. You will need a means to make the ten marks and to cut out the triangle shape (if practical). You may wish to channel energies to your talisman through a crystal grid. When you are ready, cleanse the space. You can burn sage incense, or carry out the pentagram visualization (see pages 116–117).

Create your talisman by making the ten points of the Tetractys and then cutting out the triangle shape. All the time have in your mind the intention you wish to manifest. With your hands over the talisman, state your intention out loud while visualizing a stream of bright light entering the talisman. Watch it flow from the monad at the top, illuminating each point. As the energy flows, it leaves a trail completing the implied triangles between each point. Finally, see the energy complete the tetrad before illuminating the surrounding triangle. Continue to visualize this intense energy for a time while focusing on your intention.

Seeing the Tetractys with completed triangles offers you the chance to notice how it makes nine inner triangles and one outer triangle, another manifestation of ten.

With your talisman now complete, thank the divine source. You can carry your talisman, wear it, or store it under your pillow to bring about your intention.

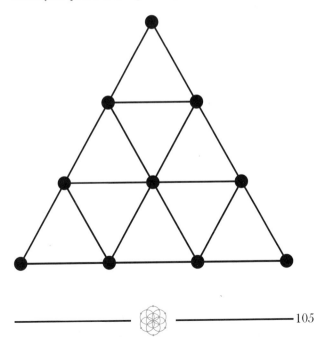

CHAKRA HEALING AND SACRED GEOMETRY

Chakra comes from the Sanskrit word meaning wheel. These are energy centres that run through the central column of the body. Each chakra vibrates with its own specific frequency, which is in turn associated with a particular colour.

There are seven chakras, and each corresponds with the colours of the spectrum from red to violet. Chakras represent a holistic view of the person as a physical, mental, emotional, and spiritual presence. Meditative practice allows us to become sensitive to these centres of energy. When the chakras are balanced and aligned, the physical, mental, emotional, and spiritual self may be said to be in harmony. Balanced chakras are therefore an important part of holistic well-being.

Each chakra may be associated with one of the five Platonic solids.

 The root chakra concerns stability and support. This resonates with the vibration of the hexahedron, the Platonic solid of earth.

The sacral chakra concerns well-being and emotion. This resonates with the vibration of the icosahedron, the Platonic solid of water.

 The solar plexus chakra concerns power and wisdom. This resonates with the vibration of the tetrahedron, the Platonic solid of fire.

 The heart chakra concerns love, forgiveness and compassion. This resonates with the vibration of the octahedron, the Platonic solid of air.

The next three chakras in the higher part of the body all resonate with the vibration of the dodecahedron, the Platonic solid of universal substance.

 The throat chakra concerns communication, wisdom and creation.

 The third-eye chakra concerns insight, perception and intuition.

 The crown chakra concerns balance and oneness with all that there is.

EXPERIENCE THE FLOWER OF LIFE

The Flower of Life is a potent geometric structure within sacred geometry. We have already seen how the Flower of Life is associated with growth, flourishing and the awareness of connectedness with all things (see page 46). We have also seen how the Flower of Life can reveal the divine Tree of Life by focusing on the central column in the design (see page 47). In this exercise, we will draw and colour a Flower of Life pattern as a meditative practice.

Many people find mindful colouring a therapeutic and relaxing activity. Reciting a mantra while creating a sacred design can help make the experience more immersive and more focused towards a spiritual intention. A mantra can be a short phrase, like a simple prayer, and may be articulated on the lips silently or repeated in the mind.

Before you start, get prepared with a pair of compasses, some art paper and colouring materials. You need everything to hand to minimize disruption. When you are settled, adopt a meditative frame of mind. Relax with some deep breaths and release any tension you are aware of.

Decide on a short phrase or word that sums up your intention for this exercise. Something simple might be, "Flower of Life, help me bloom." Repeat your phrase either out loud or in your head. It might be tricky at first, but once you get into the rhythm it can become quite natural, and you stop noticing your mantra.

Draw and colour your Flower of Life however feels best for you. Begin with the central circle using your compasses. Keep the radius the same as you draw the surrounding circles. Take your time and relax into the task. Avoid being self-critical of your work. Simply accept what happens as part of life's flowering.

When you finish, thank the Flower of Life. As your flower is infused with your mantra of intention, each time you look at your work, it will inspire you further towards your goal.

⌒ GEOMETRY JUST FOR FUN ⌒

This is an exercise to discover some amazing relationships between circles, the *vesica piscis* (see pages 34-35), and three regular polygons. There is nothing like drawing these things for yourself to discover the relationships within and to find the joy in sacred shapes. You will need:

* A pencil
* A pair of compasses
* Paper
* A ruler

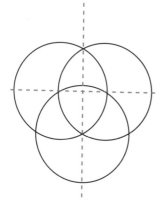

Start by drawing two circles in a *vesica piscis*. Connect their centres with a horizontal construction line extending out to their circumferences. Keep the same radius and draw a third circle at the bottom intersection of the first two. Draw a vertical construction line and extend some way above your *vesica piscis*.

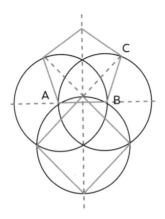

Finding the square in the bottom circle should be straightforward. Extend the upper two sides of your square to meet your upper circles. You now have the points needed for three sides of a regular pentagon. To find the final point for your pentagon, place your compass point at point A and set the radius to point C. Draw a curve to meet your vertical construction line. Do the same from point B. Now you're all good to finish your pentagon. You can finish your diagram by finding those equilateral triangles inside your *vesica piscis.*

⌒ **MORE GEOMETRY FOR FUN** ⌒

This next exercise is an opportunity to see more of the creativity that can be found in the Seed of Life pattern. The Seed of Life naturally produces a hexagon inside itself, but with a few line extensions you can see how a six-sided shape seems to organically become a 12-sided one.

Start with a Seed of Life pattern by simply drawing a central circle. Then, keeping the radius the same, place your compass point on the circumference and draw another. Use the intersection point for your next centre and keep going. With all seven circles in place, you can outline the inner hexagon.

Now all you have to do is use opposite angles in your hexagon to extend out the lines to the outer edge of the Seed of Life circles. With these lines in place, you have all the intersections you need to complete a 12-sided figure.

Notice the sacred hexagram that emerges from the centre and the straight-curve-straight perimeter of the whole shape. You have all the lines and sections you need now to design your own stained glass window like those you find in cathedrals.

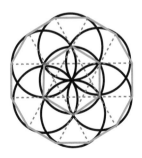

◠ **MAKE A PLATONIC SOLID** ◠

One way to experience the structure of the Platonic solids is to make them. The nets for the solids are shown below, although you can download and print nets that have the necessary tabs for gluing. You will need strong card and a craft knife to score the folds.

They are straightforward to make, with the exception perhaps of the icosahedron. Choose card of the corresponding colour (see table on page 95). Before you glue them together, you could write your own spiritual intention inside. Once you've made your Platonic solids, you could place them somewhere in your house to manifest their energies.

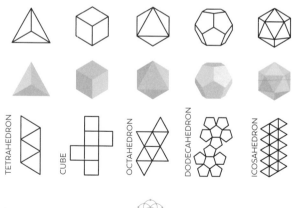

TETRAHEDRON

CUBE

OCTAHEDRON

DODECAHEDRON

ICOSAHEDRON

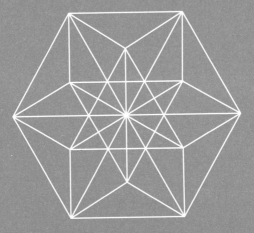

GEOMETRY IS THE ARCHETYPE OF THE BEAUTY OF THE WORLD.

JOHANNES KEPLER

A PENTAGRAM
SPACE-CLEARING
VISUALIZATION

This visualization is a useful tool to create a sacred space either as preparation for meditation or to begin a ritual exercise for manifestation. It will clear the space of any negative energies and set up some protective pentagrams. You can use this visualization as preparation before any of the meditative, manifestation or healing exercises included in this chapter. It is equally useful before tarot readings.

Stand in the centre of your working space. Facing east, with an outstretched hand and extended finger, draw what is known as a banishing pentagram in the air. Start from knee height and reach up to your crown as you draw. Visualize the pentagram as drawn in bright blue light like that of a gas flame.

Start   **Banishing**

Move to the south and complete a second pentagram. Do the same in the west and then in the north. Return to face the east.

Visualize your pentagrams as clearly and brightly as you can while they surround you. Try to sense the energy of the pentagrams and notice how the space now feels.

As you get used to this exercise, you can add extra elements. When each pentagram is complete, a god or goddess's name may be called upon as you point to its centre. Those familiar with the goddess's energies might choose Isis, Astarte, Hecate or Demeter to empower the pentagrams. If you use an athame (a ritual knife), as many Wiccans do, you can draw your pentagrams with this rather than your finger.

If you wish to draw energy into the space rather than clear it, you can draw the invoking pentagram instead.

Start

Invoking

GLAUCON: "... GEOMETRY
IS THE KNOWLEDGE OF THE
ETERNALLY EXISTENT."

SOCRATES: "THEN, MY GOOD
FRIEND, IT WOULD TEND TO
DRAW THE SOUL TO TRUTH."

PLATO, *REPUBLIC*

◠ FINAL WORD ◠

In Blake's poem, *Auguries of Innocence*, he states; "Hold infinity in the palm of your hand, And eternity in an hour." Sacred geometry is all about the immense scale of an incomprehensibly vast creation brought to the simple, the delicate and the true. This is the nature of the divine as we encounter it all around us.

This journey through sacred geometry has really been all about noticing. Appreciating sacred geometry begins with becoming aware of the patterns of life and being receptive to the wonder of it all. The patterns of life, from the Fibonacci sequence to the spiral, from divine proportions to sacred elemental solids, reveal the intricate designs at the heart of creation. These designs resonate with a living energy that seems to flow through life. Noticing these patterns, sensing connectedness and feeling this living energy can help us feel more alive and more connected with the divine source.

Next time you get the opportunity, try to take the time to notice the sacred, and know that when you hold a flower in your hand, you really are holding a piece of infinity.

⌒ **FURTHER READING** ⌒

BOOKS

Johannes Kepler (trans. C. Hardie) *The Six-Cornered Snowflake* (2014)

Gary B. Meisner *The Golden Ratio: The Divine Beauty of Mathematics* (2018)

Michael S Schneider *The Beginner's Guide to Constructing the Universe: The Mathematical Archetypes of Nature, Art, and Science* (1995)

Starhawk *The Spiral Dance* (1999)

HRH The Prince of Wales *Harmony: A Vision for Our Future* (2010)

PODCASTS

BBC Radio 4, *In Our Time*, Johannes Kepler – www.bbc.co.uk/programmes/b085xpzf

BBC Radio 4, *In Our Time*, The Fibonacci Sequence – www.bbc.co.uk/programmes/b008ct2j

BBC World Service, *Science Stories*, Kepler's Snowflake – www.bbc.co.uk/programmes/w3csy5b7

BBC Radio 4, *In Our Time*, Euclid's Elements – www.bbc.co.uk/programmes/b07881kn

WEBSITES

100 Mandalas.com –
www.100mandalas.com

Fibonacci Numbers –
www.math.net/list-of-fibonacci-numbers

The Golden Number – www.goldennumber.net

Healing Crystals Company –
www.healingcrystalsco.com/blogs/blog/crystal-grids-complete-guide

Healing Energy Tools –
www.healingenergytools.com/sacred-geometry

Museum of Science, Golden Ratio –
www.mos.org/leonardo/activities/golden-ratio

⌒ **IMAGE CREDITS** ⌒

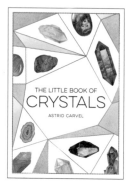

THE LITTLE BOOK OF CRYSTALS
Astrid Carvel

ISBN: 978-1-78685-959-4

Crystals have long been used for holistic healing purposes. Every crystal emits vibrations, which can help to bring balance, calm and positivity into your life. This guide will teach you how to select and maintain your crystals, along with basic techniques for crystal meditation and balancing your chakras, to bring harmony to mind, body and spirit. Discover over 40 crystals, their unique properties and how to make use of their power in everyday life.

THE LITTLE BOOK OF GODDESSES
Astrid Carvel

ISBN: 978-1-80007-198-8

Embrace the power of the divine in this beginner's guide to some of mythology's fiercest females and most legendary ladies. Learn about Athena, the Greek goddess of wisdom and war; Bastet, the Egyptian goddess of pleasure and protection; Freyja, the Norse goddess of love, and many others. You'll be inspired and empowered by the tales of feminine power, strength and wisdom of all these **dazzling deities**.

THE LITTLE BOOK OF THE OCCULT
Astrid Carvel

ISBN: 978-1-80007-722-5

Discover the wonders of occult magick

From summoning courage with a simple candle ritual, to seeing what the day has in store by reading the grounds from that first cup of coffee, occult magick has become part of daily life for many over recent years. The opportunity to glimpse the future – or learn from our past – holds a mystical allure as we try to make sense of the unpredictability of our existence. *The Little Book of the Occult* offers a fun introduction to this fascinating subject and its numerous benefits to health, wealth, love and relationships.

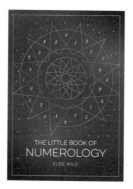

THE LITTLE BOOK OF NUMEROLOGY
Elsie Wild

ISBN: 978-1-80007-449-1

Take control of your destiny

Numbers play a significant role in all our lives, whether you have always had a "lucky number", find yourself drawn to specific dates in your calendar or hold superstitions about certain digits. This fun and informative guide will show you how to read and interpret the numerological patterns in your life so you can establish a deeper connection with yourself and others, fine-tune your intuition and make the most of life's **endless possibilities.**

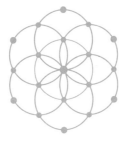

Have you enjoyed this book? If so, find us on
Facebook at Summersdale Publishers,
on Twitter at @Summersdale
and on Instagram at @summersdalebooks
and get in touch. We'd love to hear from you!

WWW.SUMMERSDALE.COM